D1017021

GOOD COMPANIONS

To my mother, Pamela Flowerdew

SUMMIT BOOKS
New York London Toro
Sydney Tokyo Singapo

GOOD COMPANIONS

A Guide to Gardening with Plants that Help Each Other

BOB FLOWERDEW

Illustrations by Sally Maltby

Summit Books
Simon & Schuster Building
Rockefeller Center
1230 Avenue of the Americas
New York, New York 10020

First published in Great Britain 1991 by
Kyle Cathie Limited

Copyright © 1991 Bob Flowerdew
Illustrations © 1991 Sally Maltby

All rights reserved including the right
of reproduction in whole or in part in
any form.

SUMMIT BOOKS and colophon are
trademarks of Simon and Schuster Inc.

10 9 8 7 6 5 4 3 2 1

Library of Congress Cataloging in
Publication Data available upon request

ISBN 0-671-75966-3

Designed by Geoff Hayes
Printed in Hong Kong

Contents

Author's Note

Most of our current knowledge of companion planting is the result of the research of Dr Ehrenfried Pfeiffer and Richard Gregg, for which I am most grateful.

It is not always possible to say *why* a particular companion effect is successful. Although experienced gardeners know that onions and beans do not prosper together, science has yet to come up with a satisfactory explanation. If we know that two plants grow well together because they like the same conditions, or because one supplies nutrients to the other, or whatever, I have said so in the text. Otherwise, we must make use of what knowledge we have gained from observation.

A note on the naming of plants: I have used common names throughout, except on the rare occasions when a plant doesn't have one. I have also normally given the Latin name of the species I am describing. If only the one-word generic name is given (as with Malus for apples), then the remarks apply to all the species of that genus.

A special thanks to Ray Rogers, who adapted the names and species for the American market.

Introduction
to
Companion
Planting

With the constant search for greater yield, size of crop or ease of production, there is one obvious question that few people have asked. Do any plants grow better or worse in the presence of any others?

The answer is, of course they do, but some interactions are so commonplace that they go unrecognized, while others may be an intrinsic part of practices many gardeners employ without realizing why. On the other hand, some companion effects are well known as such - the traditional example is growing chives under roses where they help to overcome black spot and increase the roses' perfume.

At the simplest level, plants used as hedges are beneficial, giving warmth and shelter to those nearby; bushes protect and support lilies growing through them, while French marigolds repel pests from the bedding display. Crop rotation is a hidden companion effect taking place over time. The soil condition, microlife and fertility left by one plant affects the response of the crops that follow it for several years.

A weed which needs controlling may still be a 'good companion'. Weeds protect the soil surface, accumulate scarce nutrients, convert sunlight into stored fertility and give temporary ground-level shelter to young seedlings.

There are myriad relationships continually going on unseen beneath the surface of the garden. Plants live in a state of constant interaction with pollinators, pests, predators and each other in the fight for nutrients; they are also involved in the interplay between all forms of life within the soil and above the ground.

As companion gardeners we must try to understand as much of this as possible so that we can use plants effectively to adjust and improve their environment, microclimate and soil. This includes knowing which plants to use to attract and provide food and shelter for birds, bees and all the other creatures which benefit the whole garden and beyond.

The most important thing is to be good companions ourselves. We must aim to be perfect hosts. It is essential to give each and every plant exactly the right type of soil, shelter, aspect and treatment, with extra help while it is establishing itself. This requires intimate knowledge of the plant's needs and a fair understanding of the garden's different microclimates - which spots are hot and dry, which are cool and moist, and so on.

To give optimum conditions from birth to compost heap means more than matching each plant with the best place. We must ensure that we do not handicap or harm the plant in any way, which means using gentle, organic and biodynamic techniques. With companion planting, we also consider every other plant that is, has been or will be in the immediate vicinity. We then introduce many and various plants to foster and encourage life, either directly or through intermediaries, adjacent to the plant we are concerned with and throughout the garden. Specific examples are given in the A-Z which begins on page 27.

If we succeed, plants will flourish, crops will thrive, the garden will bloom and throb with life. Fruits and vegetables will taste better, keep longer and make us healthier, while we will need to do less and less work.

How Can Plants be Good Companions?

Plants can be good companions for many different reasons. Any one relationship can be complicated by direct and indirect, negative and positive effects which all interact. For example, a tree may be companionable to a plant because the tree produces leaf exudations and root secretions that benefit the plant. The dead leaves may protect the plant's crown from frost and the tree's flowers may support a pollinator that the plant needs. At the

9

same time, the plant may be returning minerals to the soil which are essential to the health of the tree. Yet as the tree grows larger, it may sap all the moisture from the soil and its shade may kill the plant if the drought does not.

Therefore it is difficult simply to divide plants into those that help each other and those that do not. They are all frequently using the same means for different ends. Do not forget that there is no altruism in nature - each plant does what is best for it, and any good it does another plant is purely coincidental. But this does not stop us using these interactions for our own purposes. The plants will respond obligingly, if unwittingly.

So in looking at the ways in which plants can help each other, we must remember that these are not independent events. Many effects may be in conflict, and there may be interactions of which we are totally unaware.

'Direct' Good Companions

Shelter from wind and too much sun is one of the most obvious benefits plants can confer on one another. Wind-breaks and hedges are important because they curb the worst excesses of the weather, but they also warm a garden by extracting heat from the wind. Hedges reduce wind speed so that it extracts less water from the garden. As the wind slows down, it deposits debris it has been carrying, and this helps fertility.

On a smaller scale, but in exactly the same way, established plants or the stumps of the previous crop create much better conditions for young seedlings and transplants. In a garden that is sheltered from the wind, predators and pollinators can set to work earlier in the year and be active for longer each day.

Of course it is possible to overdo shelter and produce stagnant conditions or a frost pocket, though the worst problem that is likely to occur is reduced light from all the growth.

There are a number of shade-loving plants that do need sheltering from full sun, and new transplants shouldn't be over-exposed either. But for the most part plants like plenty of sun - after all, it is what makes them grow. But for those that need some protection, putting the taller plants on the sunny side will be perfect.

Many plants need cool, moist root runs. Clematis are a prime example and a companion planting of leafy ground cover will give them the conditions they crave. Spinach serves the same purpose in the vegetable garden, and has the added advantage of getting on well with all the other vegetables.

Support can also play a part in companion planting. Natural poles of sunflowers or sweet corn can be grown for peas, melons, squashes and marrows. Sturdy shrubs and trees are the best framework for climbers like roses, though more vigorous ones like *Clematis montana* may not benefit their hosts.

Beneficial root, leaf and gaseous secretions are much more difficult to observe, and they are also a lot more variable. Any secretion is affected by weather, soil and season but even more by the variety of the plant. Even within the same species there may be many different cultivars, which may or may not have the same companion effect as the original variety. It is easy to fail to observe a recorded companion effect if it does not apply to the variety you are examining! In one test carried out on 3,000 varieties of oats, only twenty-five proved to have an exudate that inhibited weeds.

The variable effects of plant secretions are inextricably tied up with plants which help each other by enriching the soil, and those which help in small amounts but hinder in large. It must be noted, though, that many plants appear to have a beneficial effect on others independent of measurable soil enrichment; the recipient probably benefits from minute amounts of hormones or by-products discarded by the donor plant.

Plants are continually passing nutrients back and forth to their root hairs, which are intermingled with the soil and are replaced throughout the growing season. Their surfaces are coated and permeated by layers of bacteria and fungi that co-exist with them. With the aid of this microlife, plants both extract nutrients from the soil solution and give it back other nutrients in exchange.

Often the microlife produce a surplus that is then available to other plants. The legumes are famous for 'fixing' nitrogen, as the nodules on their roots are very rich in this nutrient which they have obtained from the air. Nitrogen is usually only available once the root has been replaced and has died and decomposed, but root hairs are replaced more frequently, which makes for a continuous supply. Some plants, especially cereals such as barley, excrete potassium back into the soil as they ripen, and most plants put unwanted materials into leaves that are soon to be dropped.

Root channels are an invisible network that plants are constantly expanding. The root hairs are replaced regularly while the plant is living and once it is dead the larger roots start to decay as well. As the decay proceeds, the pathways driven through the soil become available to other plants and are lined with nutrients released by the breakdown. This is one reason why rotation and interplanting are so important: a different type of plant will be better able to use these 'second-hand' nutrients with less risk of disease than if they were recycled by the same species.

Enriching the Soil

Soil improvement is brought about naturally but slowly by a succession of different plants growing on a given piece of ground. Once a mat of mosses, grasses and shallow-rooting plants has stabilized the surface, deep-rooting plants like clovers, docks and thistles grow and leave pathways in the soil for tree and shrub roots to follow. We can speed up this process by choosing the plants, the order and the timing to suit the soil, the crop we want to grow and the situation.

Soil improvement may come before the crop is planted, as when green manures are dug in, or it may be produced by different plants growing simultaneously in the same plot - roses undersown with clovers or interplanted with lupins are a traditional example, though this may not be so successful with low-growing varieties. Undersown field crops benefit in the autumn when the main crop is removed, leaving them to grow on and take up the breakdown products of the decaying roots and leaf litter.

Planting the soil with different crops in rotation is like practising companion planting over time. Each crop is chosen to suit the conditions produced by the previous occupant and to leave those preferred by its successor. Plants like flax and phacelia improve the soil tilth and bind soil particles together; others, like spinach, add saponins which jell the humus together with coarser organic materials, thus helping it to hold more water.

Soil enrichment may come directly from the crop plants leaving rotting root hairs and leaf litter, or indirectly through them swelling the compost heap. The total quantity of organic material is increased by growing green manures whenever or wherever the soil is not needed by cropping plants. The quality is improved by the introduction and encouragement of the right plants to supply necessary nutrients.

Although weeds will fight to the death with our treasured plants, they are of immense value as mineral accumulators. They are capable of extracting nutrients even from soils that are extremely deficient in them. This ability is not restricted to weeds, however: Datura (thorn-apple) can accumulate phosphorus to the extent of one third of its

dry matter mineral content; the potato accumulates magnesium; and the leaves of the locust tree contain seventy-five per cent calcium. Once scarce resources such as these have been concentrated by a plant, they can be composted and re-incorporated into the soil for the benefit of other crops.

Many companion plants get on well together because competition is reduced by them choosing to root in different soil layers. The deep rooters bring up minerals from soil the shallower-growing plants cannot reach. The shallow rooters do not compete at those depths and their breakdown products will eventually wash down to feed the deeper roots.

Legumes are unique among companionable plants. They are a large family containing many garden plants which - as mentioned earlier - make available the nutrient nitrogen their companion microlife fixes from the air. Clovers are the most common legume, and the rhizobium bacteria in their roots fix nitrogen, producing a surplus which grasses can use. Planting a mixture of clovers is even more effective and improves the yield of grass much more than a single variety can.

Farmers used to grow cereals mixed with peas or beans and feed the mixed crop to animals, as separating the seed was troublesome in those unmechanized days. Tests have consistently shown that the mixture can be considerably more productive than either or both crops grown alone.

Peas and beans are equally beneficial in the vegetable garden and can be grown to advantage among many crops. Lupins, sweet peas, laburnums and several other trees are also legumes and are valuable in more ornamental areas.

The prosperity of microlife is another factor in soil enrichment: plants like yarrow and valerian encourage worms and other creatures, and the more of these there are, the more by-products and dead bodies there will be to feed the soil. Any improvement in the organic matter, saponins and scarce nutrients in the soil will encourage population explosions and thus multiply any inherent fertility the soil may have.

Insect remains are of special value to the soil life, and chitin from their body armour is believed to improve plants' resistance to disease. Plants with sticky leaves like the alder and the woodland tobacco *Nicotiana sylvestris*, or sticky stems like the German catchfly *Lychnis viscaria* not only trap many insects to benefit the compost heap, they are also useful for pest control.

Leaf litter is another valuable soil food. Plants use obsolete leaves as dumping grounds for unwanted minerals and by-products, and when they fall these leaves will feed the microlife and thus the crops. The amount of fertility introduced this way can be immense and can be increased by covering bare soil with ground cover that will trap wind-blown leaves and debris.

How Plants Oppose or Harm Weeds

In the continual battle for air, light, water and nutrients, plants have used every method available to them to oust competitors. Root, leaf and gaseous secretions are either discarded by-products or deliberately offensive chemical warfare. No plant will intentionally help another - in other words, to a plant, every other plant is a weed!

Making use of these secretions and exudations can be useful to the companion gardener, but it can also lead to complications. One early application of companion planting used wormwood to discourage cabbage butterflies. But further investigation showed that although wormwood was effective in this way, its leaves had an exudate that when it was washed on to the soil reduced the growth of the cabbages by between a quarter and a third.

Clovers are inhibited by root secretions from buttercups and vanish when the buttercups appear. This antipathy is so strong that an extract of buttercups diluted down to one part in twenty-three million still inhibits the growth of clovers! Clovers themselves give off secretions that inhibit other plants and even stop their own seed germinating at a dilution of 300 parts per million.

Many plants, especially dandelions, give off ethylene gas, which prevents seeds germinating, affects the growth of other plants and causes premature ripening. Plants that are stroked gently also respond by giving off ethylene. This then causes slower growth, which is sturdier.

The benefits that weeds can give - accumulating material for improving fertility, protecting small plants from the full force of the weather and hiding them from pests - are usually far outweighed by the competition for scarce nutrients and light, and by their persistent harbouring of pests and diseases. It is relatively easy to supply more water, mineral nutrients and even freer passage of air to compensate for those used by the weeds, but it is difficult to provide more light.

Severe shade is one of the chief causes of poor growth. When it is combined with dry conditions brought about by overhanging boughs, few plants will prosper. The length of the 'canopy period' is important: trees surrounding a garden are going to be detrimental if, like peach, they come into leaf early and do not drop their leaves until late autumn.

Trees which leaf late, like ash, are preferable, because they allow the particularly important spring light through their branches to plants below.

Some plants are more adept than others at shading out their opposition - their success can even be measured with the aid of a light meter. Taller plants are obviously at an advantage, but the density of leaves is also important in determining the degree of shade.

Oats and rye increase their shading much faster

than wheat, but neither is as effective as hemp. This now illegal crop was once popular for its fibre, and also because it smothered out weeds by creating heavy, undiminished shade.

Tagetes minuta - the Mexican marigold - is another, more useful, smother crop which has even suppressed pernicious weeds like equisetum (horsetails) and bindweed. Its smothering effect is augmented by root secretions that inhibit weeds and herbaceous plants but have little effect on woody ones. These secretions also kill the minute parasitic worms called nematodes.

Once established, shade-making plants inhibit weeds very effectively and so ground cover is useful in the ornamental garden. Dense growth, especially that of evergreens, suppresses weeds almost completely. Conifers are often effective as the exudates of their leaves or needles also inhibit the germination of seeds. Few weeds germinate under rosemary, wormwood or rue for the same reason.

In the productive garden, close planting, block planting and intercropping with companion plants can exclude light from the soil and similarly handicap weeds. The danger with these methods is that too many plants will be crammed in together and will suffer as a result. Do not forget that although they may not take up all the space above ground, there may not be enough space underground for all the root systems. The roots of ornamentals will tolerate more competition than those of vegetables.

If a plant cannot shade its rivals out, it may compete by depriving them of essential nutrients. The plants that start into growth earliest in the season can grab every bit of nutrient as it becomes available, and so the competition is starved before it starts. Perennials with massive food reserves in their roots have mastered this method of beating the opposition. They use their reserves to grow before enough light is available, which means they

take all the nutrients and light, leaving none for their competitors. Herbaceous plants also tend to use this method, but are beaten by creeping weeds like ground elder.

The elimination of nutrients is a factor in the effect known as 'replant disease'. This is not a disease as such, but is poor growth symptomatic of a plant put in or near where one of the same genus has recently been. Replant disease is worst in woody plants, especially roses and apples, but occurs to a lesser degree with most plants.

Annuals and short-lived crops are rotated to counteract this problem. Trees are obviously more difficult to rotate, and if one dies or is blown down, the temptation to put in an exact replacement must be resisted: it will never do as well as a different genus.

Replant disease is caused partly by the previous occupant of the site having used up all the nutrients suited to the genus, partly by a build-up of pests and diseases, but frequently by a residue of allelopathic chemicals, which many plants use to suppress temporarily the germination of their own seeds. (Some plants, notably the black walnut, suppress almost everything.)

Allelopathic compounds will prevent a replacement plant establishing itself for several years. By then they will have broken down, the soil will have recovered its nutrient balance and the pests and diseases which affect that genus will have dissipated, making the site suitable once more.

Using a Little of One Plant to Help Others

Many companion planting observations note that very small numbers of some plants increase yield or improve quality in nearby crops. If the ratio rises, then these good companions start to compete with the crops and yields drop. The gain is probably due to hormones or similar chemicals affecting the growth and nutrient uptake of the main planting. The benefit is then overwhelmed by the depression in growth caused by direct competition.

This phenomenon is particularly noticeable with cereal crops. Yields of wheat are improved by very small numbers of corncockles, chamomile or white mustard plants, but larger quantities rapidly choke the crop. The pungent herbs used as border plants aid most of the vegetables, but again, plant too many and competition sets in.

Stinging nettles present us with another quandary. They are hardly pleasant companions, but their presence makes fruit - particularly tomatoes - ripen fully with less rotting, and increases the oil content of herbs grown nearby.

Repelling or Attracting Harmful Insects

Plants can be used to help other plants in many ways when pests need controlling: they can hide them with strong odours or camouflage them with leaf patterns or the 'wrong' flowers. The classic example is growing carrots and onions together to confuse their respective flies.

Strong odours may not just mask crops. The smell of wormwood will actually drive pests away. Marigolds in the greenhouse repel whitefly, but will not drive them out once they are in.

Some plants have leaf or root exudations that make them distasteful to pests and they can transmit their immunity to plants growing with them by proximity or absorption. Nasturtiums are grown around apple trees to drive away the woolly aphis. Pungent herbs have a similar effect, bolstering many plants against pests.

Conversely, companions may be of benefit by being more attractive to the pest and enticing it away. 'Trap crops' lure pests on to plants that will

be destroyed, taking the pests with them. Wireworms in new soil can be attracted on to a mustard crop that is turned in before the main crop is planted. Potatoes have been used in a similar way to clear the ground for hops. Whitefly in the greenhouse are attracted to the sweet tobacco plants *Nicotiana alata* and *N. affine* with which they can be dispatched to the bonfire. Red spider mites cannot resist broad-bean plants, but take longer to move on to them.

Companions may be 'sacrificial', in which case the pest damages them rather than the valuable crop. An extreme example is to keep birds off gooseberries by planting some strawberries and raspberries near them. Blackberries in the hedges will similarly reduce damage to other fruits in the autumn. Slugs will eat lettuces, spinach and other soft seedlings before they attack tougher plants, so extra can be sown as cannon fodder. Carrot blossom moths will go to parsnips in flower and leave the carrots alone.

Sacrificial crops are also of great benefit in breeding up predators to control pests elsewhere. Aphids come in varieties that are specific to a few plants, while the predators that feed on them do not usually mind what type of aphids they are eating. Growing more honeysuckles, lupins and cherries that regularly get covered in aphids is not going to infect many other plants, but the ladybirds and other predators bred up on these will seek out any other aphids they can elsewhere in the garden.

Attracting Useful Insects

This is an area in which companion planting has much to offer, as increasing the numbers of pollinators, predators and parasites in the garden does more than just help pollination and pest control. These creatures also become food resources for hyper-parasites and insectivores, which in turn feed birds and carnivores.

All these chains of life are concentrating nutrients and depositing them as by-products and dead bodies, which build up the general level of fertility. A bees' nest may contain 50,000 bees at midsummer. None of them were alive last year and none will be next. They are continually being replaced, so a single nest can drop 50,000 to 100,000 dead bees in the surrounding area, and each body adds an extra drop of fertility.

Plants that are especially good at attracting and feeding pollinators and predators include buckwheat, the poached egg plant *Limnanthes douglassii*, alpine strawberries and *Convolvulus tricolor*. Many flowering plants will attract hoverflies and bees, but different ones appeal to butterflies and moths. Bees see in the ultra-violet spectrum and tend to prefer flat flowers with bluish colours. Butterflies use scent and pheromones - sex scents - more, so colour is less important; they go for orange and red shades with deep, tubular flowers.

A garden containing a wide range of plants will bring in and encourage the many and various pollinators - not only bees and butterflies, but also beetles and flies. In the same way, the populations of many different predators and parasites can be increased. The aim is to provide a staggering variety of food (flowers and pests) and shelter over as long a season as possible so that there are always plenty of useful insects about.

Repelling Animal Pests

Plants that have thorns and prickles are excellent deterrents and good companions in the garden. They keep dogs and cats from trespassing, trampling plants or fouling the soil. On the other hand, they can be annoying to humans!

Some plants give off smells that are repulsive to animals. The leaves of elder are said to discourage

mice, and rabbits are supposed to dislike onions. Root exudations from euphorbias are an ever-present possibility for a reliable mole deterrent, while the pungent herbs and the leaves of plants such as wormwood can be used to discourage slugs and snails.

Repelling or Reducing Fungal and Other Plant Diseases

Most fungi are saprophytic and live on dead material which they decompose, but many also attack living plants and creatures. Bacteria are similarly most often concerned with dead sources of food, but there are plenty of varieties that attack living cells.

Many root exudations and plant secretions act as chemical warfare against these attackers, and they in turn use chemicals to gain entry to plant cells. Fungi also use their own secretions to destroy bacteria. We have gained most of our antibiotics, including penicillin, from research into these chemicals.

Little is known about how different plants may aid others against disease, except that anything that promotes plant health is naturally going to help prevent disease. Plants containing a lot of silica offer most benefit, and sprays made from garlic, stinging nettles and equisetum have proved effective.

Good Companions for Animals and People

Food plants are obviously beneficial, but here we are referring to aesthetic effects like scent, and also to medicinal uses. Many plants are a delight to the human senses, and it is rather arrogant of us to believe that other animals cannot appreciate them at all. A beautiful plant, a wonderful scent, pleasant surroundings generally, are very good for restoring physical and mental health and that is, after all, the reason most of us have gardens anyway.

However, many plants, particularly the herbs, are attractive to animals for medicinal reasons. Natural medicines occur in a wide variety of plants, and animals will find these to cure themselves: we have all seen cats and dogs eating grass. The beneficial effects on animals and people are indisputable, and over half our modern medicines are still based on plant products. It is sad to reflect that if we all ate more fresh, organically grown food we would probably not need most of these potions in the first place.

Attracting Birds and Other Creatures

Although birds do a lot of damage, their overall effect is beneficial, especially as they concentrate fertility and add life to the garden. Birds are encouraged by those plants offering perches, shelter and food. If you have a lot of plants providing flowers, fruits and berries over as long a season as possible, you will build up a population of birds that will control pests and eat many weed seeds.

Plants giving good ground cover and cool, moist shade encourage frogs, toads, newts and slow-worms which all eat pests (though to be honest they will also eat any other beneficial predators too slow to escape) and add their bit to the garden's fertility.

The Development of
Garden Companions

Directly or indirectly, we can take advantage of all these ways in which plants are interacting. Plants adapt naturally to the conditions around them, including other plants growing nearby. When the conditions change too much for a plant to adapt to them comfortably, another plant will take over, simply outcompeting the original by being more suited to the area. Any plant that has been able to flourish and outgrow weeds, pests or diseases for many years - or even centuries - is by definition a survivor and will get on well with the other chosen plants in the garden.

The cultivars that go together to make up our gardens have been selected in many ways from an immensely larger number of potential and un-likely plants, sometimes deliberately but more often by chance. New plants, with or without some obvious usefulness, are still being discovered, but the vast majority were recorded and catalogued by the end of the last century and many have been with us since long before Roman times.

Initially, plants would have been selected for yield, colour or usefulness. Their size, physical form and nature then dictated how they should be treated. Early cultivation must frequently have been a bit hit and miss, so it is sadly likely that trial-and-error experimentation discarded many hope-fuls we could have enjoyed today.

Centuries of cultivation have altered the candi-dates that were originally most suitable, almost beyond recognition in some cases. The members of the modern cabbage family are very different to the wild coleworts, carrots have changed colour over the years, and celery was once very bitter.

All the whims and vagaries of fashion, accidents of history, varying soils and climates have contrib-uted to changing and reducing the vast numbers of possible plants to those we can choose from today.

However, the choice still runs into many tens of thousands of different cultivars for food crops, and even more for ornamental plants.

Gardeners have found it practical to split this multitude into more manageable groups that tend to get on well together. This grouping has been put into practice and formalized in the design of gardens, with separate areas for vegetables, shrubs, alpines, herbs, the herbaceous border and the orchard.

These groupings may sometimes be purely for convenience, but most of them show the important relationships that exist between certain plants. Within each group, the majority of plants do have much in common, preferring similar treatment and conditions. Moreover, many of them have been selected and grown together in these groupings for as long as they have been known and cultivated, and so have become used to each other.

So companion plants already exist in groupings all around our gardens. Hundreds of years of culture have ensured that only those that do get on with the others have been propagated: any 'anti-social' plants will rapidly have been evicted or placed away from the rest. These groupings continue to consolidate as most of the plants not only prefer the same conditions, but are continuously selected and adapted to do so.

We carry on using these groupings from habit as much as anything, and not always with the best results. The fruit cage has come to include a selection of woodland currants and berries which all get on together, but we rarely think of planting the ground cover they need to keep the soil moist.

On the other hand, convention has dictated that vegetables are rarely put in the shrub border. There is a current fashion for introducing food plants into the flower garden, but this rarely proves satisfactory. Although one of the purposes of companion planting is to reduce monocultures and increase the diversity of

plants, it does not work if their individual needs are not met. Vegetables in the flower garden can occasionally look attractive, but if you eat them you will leave an unsightly gap. What is worse is that many vegetables just will not prosper amongst such varied and efficient competitors, though fruit trees, culinary herbs and berries are easier to establish.

Now I am not suggesting that it is impossible to make a 'flower' garden that is primarily aesthetic and also productive. It's just that it must be planned from the start. The choice of crops is slightly limited and 'all-year-round interest' difficult to maintain, but good planning can compensate. What cannot be done successfully is introducing most food crops into an established flower garden: they simply will not 'do'.

The romantic picture of the cottage garden is a gross misconception. It was never a major contributor of food. Cottage gardens were full of herbs for medicinal use, flowers for pleasure and fruit trees. The bulk of the cottager's diet was beer and bread with the addition of some field crops such as turnips, and many types of wild produce. Do not be misled either into thinking that these pretty gardens needed little work. Plants of many different varieties closely planted will only look good with clever design, regular attention and almost continuous hand weeding in the spring, though the good news is that they will not suffer badly from pests.

The converse, putting flowers in the productive garden, is a much better arrangement. Give priority to the productive plants, making sure they have the best soil and situation, then introduce the ornamentals to bring in their benefits as companions and to add beauty.

Eventually the whole garden should become, and be seen as, productive, not just of fruit and vegetables but of material for increasing fertility and the active promotion of all forms of life.

A-Z of
Companion
Planting

Alfalfa
(Lucerne)
Medicago sativa

A leguminous perennial that
roots deeply, accumulating
iron, magnesium, phosphorus
and potassium, as well as fixing
nitrogen. It is excellent for
grassed areas with deep sward
like orchards, keeping the
grass green longer in droughts,
as it can reach moisture and
minerals inaccessible to more
shallowly rooted grasses.

Alliums

These members of the onion
family, related to the lilies, form
bulbs, most of which are used for
culinary purposes - onions, leeks,
shallots, garlic and chives. They
have a strong smell which can
help hide other plants from pests;
as they accumulate sulphur, they
also have a long-term fungicidal
effect. Chives are often grown in
a ring under fruit trees and roses
for this purpose, hiding the lower
stems into the bargain. Reputed
to repel moles (you should be so
lucky), they are among the most
effective companions for many
plants, especially carrots, but
should be kept well away
from any of the beans.

Anise
Pimpinella anisum

A pungent annual herb used in
cooking and in ointments to protect
against insect bites and stings.
Germination, seed formation and
vigour of most plants improve when
sown with coriander. Anise is a good
host plant, giving nectar to predatory
wasps which deter aphids. Its scent
hides brassicas from pests by disguising their distinctive smell.

amaranth

28

sweet corn

Amaranth
Amaranthus retroflexus

This unusual but highly
nutritious tropical annual is
used as a salading or a
substitute for spinach. It
needs hotter conditions than
are found outside in the UK,
but given these it will do
well with sweet corn as its
leafy shade provides a rich,
moist root run; this also
makes it an excellent host
plant to ground beetles.

Apple
Malus

Apples are grown everywhere so are vulnerable to many pests and diseases. They are protected against scab by alliums, especially garlic or chives, planted in a ring underneath. Nasturtiums, particularly the trailing varieties, discourage American apple blight or woolly aphis if they are grown near apples every year, so that the trees absorb the repellent flavour of the nasturtiums. Lay apples on dried nettles to preserve and ripen them.

apple

allium

Artemisia

This genus contains many plants with insecticidal properties. Wormwood is the strongest and most effective; it gives off an unpleasant smell and has poisonous exudates. Southernwood is friendlier and delightfully scented; mugwort is the wild form and tarragon the one we use in the kitchen. As southernwood is so delightful to brush against, it is pleasant to plant by paths and gateways and the scent will then hide crops from pests.

Asparagus
Asparagus officinalis

Traditionally grown within the vineyards of Beaujolais as it likes the conditions under vines and the berries fob off the birds, asparagus grows well with tomatoes, which will protect it from the asparagus beetle; in return a root secretion from the asparagus kills trichodorus, a nematode that attacks tomato roots. Both plants like basil and the three grow well together if conditions are warm enough. Parsley does well with asparagus, but only if both can get the moisture they thrive on.

Aubergine (Eggplant)
Solanum melongena

Related to tomatoes, potatoes and peppers, this is often grown with the latter as they like the same warm, rich conditions. It grows well with peas, tarragon and thyme and you can protect it from Colorado beetles by interplanting it with beans.

Basil
Ocimum basilicum

One of the tastiest herbs in the garden – it is a shame it needs so much warmth. It's delicious to eat with tomatoes and also grows well with them; they make a happy trio with asparagus. The purple, small-leaved and lemon-scented varieties of basil are also divine flavourings which should be grown more often. Because they are more attractive to aphids than many crops, they can be used as trap plants.

Bay
Laurus nobilis

Often grown in pots, this makes a small tree if not cut back by wind and frost. As a bush it can give year-round shelter to other, more tender herbs. The leaves preserve grains and seeds from weevils and the burning leaves are poisonous to humans and to insects.

Beans

All beans are leguminous and should be grown more a
they enrich the soil with nitrogen fixed from the air. I
general they do not thrive with the onion family but d
well mutually with carrots, brassicas and beets; they al
help cucumbers. They may be grown with borage,
potatoes, squashes, strawberries or tomatoes and are a
great boon to sweet corn and cereals, providing the
nitrogen these crave. In small portions they may aid lee
and celeriac. In the USA marigolds will protect bean:
against the Mexican bean beetle, as will potatoes.

broad bean

Beans, Broad
Vicia faba

Often autumn-sown to avoid the black aphid,
which may otherwise be discouraged by the sm
of summer savory; this also cooks well with beans. Bro:
beans are a good early crop to precede maize, as
they leave a rich, moist soil and give wind protection to
the young shoots. Similarly they can come before and
overlap with the brassicas (cabbage family), which can be
interplanted amongst the ripening beans to take up their
breakdown products as the roots decay. Beans and potatoes
can be sown together in the autumn for earlier crops if the
winter is mild. The seedling beans protect the early potato
shoots from wind and frost. Planted amongst gooseberries
they discourage the gooseberry sawfly caterpillars.
Field crops of beans smell sweet and
provide a lot of nectar for bees.

Beans, French, Haricot
Phaseolus vulgaris

These originated in South America and are known as wax-pod, snap,
string or green beans. There is even a running French bean. They do
well with celery in rich moist soil and produce earlier crops when planted
with strawberries, which also benefit from the supply of nitrogen these
legumes leave in the soil. Sweet corn is a good companion with French
beans and both get on well with the cucumber family. French beans,
sweet corn and melons are also a good combination, requiring warmth
and shelter which the others can help provide.

haricot bean

32

Beans, Runner
Phaseolus coccineus

These have the same
affinities as the other
beans and get on
with most plants
except the onion family.
As they are climbers
they can be grown up
and over maize or
sweet corn,
protecting them
from corn army-
worms. They thrive
alongside most
brassicas,
especially Brussels
sprouts which are
sheltered while small
and then can grow on
once the beans die back,
flourishing on the nitrogen
left by the beans' roots.
Late in the season runner
beans tend to shade out
other crops; this may benefit
some, especially celery and
salad vegetables, but only
when enough water is
available.

runner bean

33

Bee Plants

senecio

Bees are essential to many crops as without them there would be little pollination. Honey bees are really worth encouraging and are easy to keep in the garden; bumble bees are even more valuable because they fly in colder conditions and will pollinate earlier in the year than the honey bees – but, of course, they give no honey. They both need continuous supplies of pollen and nectar, so many different flowers should be grown to spread over a long season. Leaf-cutter bees are important for pollinating early fruit; they benefit from the same plants as the others.

Honey bees go for bluish or whitish flowers but other colours are not excluded. The following garden 'flower families', commonly grown in beds and borders, have built up a close relationship with bees and will attract them; those marked with an * are some of the best for putting amongst the vegetables or in borders to keep the bees visiting the area: Aconitum, Allium*, Anchusa, Arabis*, Aster, Borago*, Campanula, Centaurea, Chionodoxa, Colchicum, Delphinium, Echinops, Echium, Endymion, Erigeron, Hyssopus*, Lavandula*, Limnanthes*, Limonium, Lobelia, Lychnis, Lysimachia, Lythrum, Malva, Matricaria*, Melissa*, Mentha, Monarda*, Muscari, Myositis, Nemophilia, Nepeta*, Nigella, Omphalodes, Origanum*, Papaver, Phacelia, Platycodon, Polemonium, Pulmonaria, Reseda, Scabiosa, Scilla, Thymus*, Veronica.

Trees and shrubs have also come to depend on bees. Most members of the following 'families' are good for bees, but avoid sterile, double-flowered varieties: Acer, Aesculus, Alnus, Berberis, Betula, Caragana, Catalpa, Ceanothus,

Cercis, Chaenomeles, Cistus, Cotoneaster, Crataegus, Daphne, Elaeagnus, Escallonia, Fagus, Fraxinus, Fuchsia, Hedera, Hypericum, Ilex, Laurus, Liquid-ambar, Liriodendron, Malus, Mespilus, Olearia, Perovskia, Rosmarinus*, Rubus, Salix, Salvia*, Senecio, Skimmia, Sorbus, Spiraea, Symphoricarpus, Syringa, Tamarix, Tilia (the common lime or linden is well known for filling hives with honey), Ulex, Viburnum and Weigela.

In the vegetable and fruit garden, leave unwanted brassicas, leeks and onions to go to flower, as bees love these. Strawberries, especially the alpine ones, are always popular, but raspberries, blackberries and the hybrids such as tayberries and loganberries are favourite. Clovers sown in long grass and as ground cover or green manure are one of the biggest yielders of pollen; on farms field bean and oilseed rape (canola) are the largest sources. Sweet basil, summer savory, lemon balm and the mints are all much loved. For cost effectiveness over large areas, mints are probably the best as they spread so well.

Weeds are, of course, by definition, mostly unwanted, but the flowers of these are loved by bees: agrimony, bellflower, betony, corncockle, clovers, cranesbill, flax, forget-me-not, hound's tongue (cynoglossum), mallow, meadowsweet, melilot (American sweet clover), mullein, ox-eye daisy, pansy, poppy, ragged robin, scabious, soapwort, tansy, teasel, trefoil, valerian, viper's bugloss and yarrow.

In early spring, bumble bees prefer aubretia, berberis, bluebells, dandelions, flowering currants, wallflowers and white deadnettle. From early summer they go for brambles and their hybrids, buddleias, clover, comfrey, cotoneasters, fuchsias, globe artichokes and cardoons, goldenrod, jasmines, knapweed, lavenders, mallows, Michaelmas daisies, rhododendron, thistles and vetches.

A hedge or tree covered in ivy will provide late autumn flowers full of pollen to help any weak colonies build up stores to overwinter; surround a bee garden with an ivy wind-break to gain a double advantage.

Beech
Fagus sylvatica

These are big trees, and not good companions for the small garden! They improve the fertility of the soil and beech-mast or seed is valuable to wildlife. The leaves were traditionally best for mattresses.

Beets
Beta vulgaris

Many associated plants make up this genus. Red, yellow and sugar beet are very closely related to leaf beet, perpetual spinach and Swiss chard. They are all good mineral accumulators; one quarter of the mineral content of their leaves is magnesium, which is extremely valuable when added to compost. They do well growing with most beans, though not with runners. They like to grow with lettuce, onions and brassicas, especially kohlrabi. Birds love the leaves of all the beet family and growing them under beans or brassicas will help hide them while they are small and tender.

Birch
Betula alba

The peeling bark is impregnated with resin that has insecticidal properties and is gathered by birds for nest building. Birch roots give off secretions that accelerate composting so they may be planted as screens for manure or compost bins. Their presence stimulates growth in larches.

beech

Bird Plants

Gardeners often wish to discourage birds, but on the whole they are beneficial - most birds control pests more than they spoil crops. The common sparrow and pigeon are exceptions. Individual crops may need protecting if birds are encouraged, but we all benefit in the end from more birds. The right habitat and food will help attract them, but water is essential and like us birds appreciate a bath - where the cat can't get at them. As is soon apparent, birds much prefer our cultivated fruit to the wild fare nature provides. Often they are just after the succulence, which means the damage can be reduced by providing water for them to drink. The more fruit you grow the less harm your resident population can

do, so plant extra sacrificial fruit trees. These are some of the most bird-friendly trees and shrubs:

Amelanchiers are very pretty and covered in small fruits from midsummer.

Beech is excellent as a hedge: it provides cover for nests and perches. It keeps its leaves during the winter, giving good shelter.

Berberis (barberry) is a prickly genus with spiky leaves and thorns giving protection, shelter and plentiful berries.

Conifers provide excellent nest-sites, roosts and protection against the weather.

Cornus mas, Cornelian cherry, has an edible fruit following yellow flowers.

elderberry

Crataegus or thorns are beloved for nest-sites and berries.
Elaeagnus pungens is a dense evergreen which provides good nesting-places and weather protection.
Elderberry is one of the very best for perches, nests and berries.
Holly is well known for dense, evergreen growth and occasional berries.
Laurel is better for shelter than for nesting, as it is easily climbed by predators.
Lonicera or honeysuckle plants are not all climbers; some, like *L. purpussii* and *L. fragrantissima*, are shrubby and winter-flowering. The dense growth makes them good for nests and they produce berries.
Mahonias are related to the berberis family and have holly-like leaves; the grape-like fruit is edible and birds love it.
Pyracantha or firethorn has horrible spines and loads of berries.
Shrub roses are impenetrable to preda-tors and produce lots of hips.

Sorbus, the rowans (mountain ashes) and whitebeam, are renowned for plentiful berries.
Yew is one of the best nesting trees as the vertical trunks are full of nooks and crannies hidden by ubiquitous evergreen shoots; the berries are loved by birds but are poisonous to humans.

Climbers which attract birds include the following:
Ceanothus is a not very hardy group of blue-flowered shrubs that need a wall for protection as much as support. They produce dense growth and are mostly evergreen, which makes them good nest-sites.
Clematis romp everywhere and nests can be found all through them. The seeds of many have fine floss that can be used for nesting material.

pyracantha

Fruits loved by birds include the following:

Apples are much appreciated, so plant as many of them as you can squeeze in and the fruit will all be gone by the new year.

Blackberries are traditional nesting-places; the cultivated berries yield far more than the wild.

Blackcurrants are not as popular with birds as redcurrants, but the berries still go.

Cherries disappear with tremendous speed and the large trees make good perching- and nesting-places.

Gooseberries are eaten despite the thorns.

Grapes are good at producing berries and nesting-sites, especially if they are allowed to ramble.

Loganberries are rarely seen ripe if you leave them unprotected.

Mulberries are slow to fruit, so it would be far-sighted and worthwhile to plant them for birds to eat while you rest underneath them in your old age.

Peaches may go apparently unrecognized on the trees for years, but as soon as one bird discovers them they all learn.

Pears are rarely allowed to ripen on the trees, let alone rot on the ground.

Plums are on the ends of twigs and sustain less damage than most fruit until they hit the ground.

Raspberries are one of the crops birds love most; the autumn-fruiting ones extend the season.

Redcurrants are strongly affected by a fruit cage; you get 10,000 per cent increase in yield. In other words, without a fruit cage you are left with almost none!

Strawberries are *the* crop for birds; they cannot get enough so plant many varieties and include the alpines to extend the fruiting season.

blackberry

Tayberries are a new, improved, more bird-palatable type of loganberry.

Whitecurrants are versions of redcurrants that disappear nearly as fast.

Worcesterberries have dense, thorny growth and are like gooseberries, but are larger and most suitable for wild corners.

Birds also love vegetables. If the fruit fails to satisfy the birds' appetites you can sow some of the following vegetables for tasty little seedling snacks:

Beet and their family are razored off as soon as they emerge. This includes red and yellow beetroot, Swiss chard and leaf beet.

Brassicas are much liked by pigeons. Small leaves are attacked but once established they become too tough for garden birds. Chickens love greens and if you let them loose among brassicas, they choose cauliflower and broccoli leaves first.

Lettuce are a ready meal for birds until bolting, when they become bitter - although chickens will eat them at this stage.

Peas are another little snack much eaten when seedlings are small.

Spinach rarely survives emergence unless protected.

whitecurrant

cherry

borage

Borage
Borago officinalis

One of the best bee plants, this is a good accumulator of minerals for compost and grows well with strawberries. It is also used to repel attacks of Japanese beetles and tomato hornworms.

strawberry

Blackberries and their hybrids
Rubus

These bushes are sanctuaries in nature - small birds and creatures can hide inside, protected by the arching thorns. The centres are often bone dry and provide a good place to nest or hibernate. The late flowering caters for insects in early autumn when food is getting short and the fruit then feeds birds till winter. They may aid grapevines and certainly provide a sacrificial crop at the same time as the grapes ripen. They may be helped by tansy and stinging nettles.

blackberry

brassica

dill

Brassica

The members of the cabbage family need rich
soil and plenty of lime. They benefit from herbs such
as chamomile, dill, peppermint, rosemary and sage which
can all be planted in the borders around the vegetable beds.
Brassicas do well with peas, celery, potatoes, onions and
dwarf beans. A good trio can be made if the peas are grown
up the middle of the bed and flanked with potato plants
and brassicas. If the potatoes are wide-spaced to get larger
tubers, they can be alternated with the overwintering
brassicas. As the peas and potatoes are cleared, the
brassicas grow away, taking up the breakdown products as
the old roots decay.

44

Broccoli
Brassica oleracea

Winter cauliflowers are really broccolis as they are hardier than true cauliflowers. They are highly bred and require very rich conditions and heavy soil to form the swollen and immature lateral and terminal flower-buds we love to eat.

Subject to the same problems as the other brassicas, they need lots of care. The shape and texture of their heads make pest problems more detrimental than for, say, cabbage, where the outer leaves can be discarded. Interplant them in a trio as for the cabbages.

Brussels Sprouts
Brassica oleracea var
Bullata gemmifera

One of the hardier brassicas, they have the family tendencies, differing mainly in a greater tolerance for runner beans and a liking for really firm soil. They can be planted into an onion bed as it matures and will benefit from the onions' root residues and packed soil.

rosemary

Buckwheat
Fagopyrum esculentum

One of the best hoverfly atttractants, this is a calcium accumulator and useful green manure, and should be grown more often. With its pretty pink flowers but rather straggly growth it is probably best tucked away out of sight at the back of borders or beds.

Butterfly Plants

Butterflies are not pest controllers and
may themselves be pests in disguise.
However, they may pollinate and their
beauty is reason enough to encourage
them. Most flowers can feed them as
they have long tongues (of course many
rely on specific plants for the caterpillar
stage). The scent of the flowers mimics
the sex pheromones of butterflies and
moths to attract them as pollinators. In
the dark this is obviously more effective
than colour and so accounts for the many
luscious evening and night
scented plants.

sedum

The Buddleias, *Sedum spectabile*, goldenrod, valerian and lavender are among the best attractants, also Dianthus, Hesperis, Hyssop, lilacs, Loniceras, Lychnis, Lythrus, Myositis, Oreganum and Violas.

buddleia

Cabbage
Brassica oleraceae capitata

The commonest brassicas, these are terminal buds and to get them to swell without opening is a marvel of controlling nature. Constant unchecked growth in rich, moist conditions is required and can be aided and protected by strong herbs. Grow dill, mints, rosemary, sage, thyme, hyssop and chamomile nearby in the borders; cabbages will also be happy growing in among a bed of onions or garlic. As these all get along with beets and chards, you can make a useful trio. Alternatively broad beans, cabbages and the beets can be interplanted. Peas, potatoes and cabbages will make another successful threesome.

catnip

48

Caraway
Carum carvi

This is deep rooted and difficult to establish. The suggestion has been made that caraway and peas can be sown together. The caraway germinates once the peas are cleared and the surface disturbed. I suspect this only works with a long growing season!

Carrot
Daucus carota

Plagued by the carrot root fly, numerous herbs and strong-smelling remedies, mostly from onions, leeks and salsify, have been used with some success. Carrots' general health is better near chives, celery, lettuce, radish and tomatoes. They promote growth in peas but dislike anise and dill, although the latter discourage the root fly. If left to flower, carrots attract hoverflies and beneficial wasps.

Catnip
Nepeta cataria not mussinnii

Useful for driving away fleas and ants if you can stand the smell yourself. Most cats love it, so it can be planted where you want them to lurk; a patch near the strawberries can really keep down the bird damage. Of course if you are bothered by cats on your warm patio, then put a bed of catnip in a hidden sunny spot and they will soon gravitate there. The scent helps repel aphids, flea beetles and in the USA Colorado beetles, darkling beetles, Japanese beetles, squash bugs and weevils. It is very attractive to bees.

Cauliflower
Brassica oleraceae

Treat as broccoli, above.

strawberry

Celeriac

Apium graveolens rapaceum

This is much easier to grow than celery but still needs constant moisture. It does best in rich soil following legumes, especially after a green manure of vetch. Celeriac will grow well between rows of runner beans if kept moist - and spraying helps the bean flowers set. It is a good companion to grow with leeks, which can be tolerated by the runner beans, but these plants will only make a trio if that essential moisture is abundant. Celeriac will similarly do well with most brassicas, onions and tomatoes.

chervil

Chamomile, German

Matricaria chamomilla not
Anthemis nobilis or *cotula*

Cabbages and onions improve both in yield and flavour if there are a few chamomile plants nearby. Chamomile accumulates potassium, calcium and sulphur, which are later returned to the soil; it is also host to hoverflies and wasps. *Matricaria chamomilla* has traditional medicinal as well as companion effects. It increases oil production from plants like peppermint and stimulates composting and yeasts; the extract increases the growth of yeasts even when diluted to one part in eight million.

Celery
Apium graveolens

Celery does well in the shade of beans and tomatoes and benefits brassicas by deterring cabbage-white butterflies. It grows best of all with leeks. If left to flower, the celery and leeks will attract many beneficial insects, especially predatory wasps.

Chervil
Anthriscus cerefolium

An annual herb that likes the shade and moisture. It aids radishes and keeps aphids off lettuces.

radish

51

Chinese Cabbage
Brassica pekinensis

This is eaten first by flea beetles and slugs and
then by aphids once the inevitable bolting sets in,
so it makes a good sacrificial or trap crop. It also
grows well in the shade of Brussels sprouts. In the
USA it is used as a sacrificial crop to maize as it
attracts corn-worms.

Chives
Allium schoenoprasum

One of the Allium tribe and
probably the best to grow to
keep fungal diseases
down. Use it
against black spot
on roses and scab
on apples - but be patient,
as it will be three years before
it takes effect. Chives
discourage aphids on chrysan-
themums, sunflowers and
tomatoes and benefit carrots.
Chive sprays have been used
against downy and powdery
mildew on cucumbers and gooseberries.
As they are beneficial to us as well, they can be
planted in many places to ensure a
continuous supply.

Chrysanthemums

These are members of the daisy family (*Compositae*) and many of
them produce exudations from leaves and roots which protect
against pests and diseases. *C. coccineum* is effective at killing
nematodes while its flowers and those of *C. cinerariaefolium* have
been used as insect killers for 2000 years. Their insecticidal
properties are useful in all parts of the garden.

chive

chrysanthemum

Clover
Trifolium

Leguminous, low-growing plants that attract many insects, especially bees, clovers are among the best short-term ground covers, green manures and plant companions. Sow sweet white or red clover as companions to orchards; always encourage them in grass, as this keeps the turf green longer in dry weather. Clovers provide good cover for ground beetles and are host plants to predators of woolly aphis, thus helping apple trees. They will deter cabbage root fly if sown underneath.

Collards
Brassica

These are primitive cabbage plants grown for their heads. In hot conditions they suffer greatly from flea beetles. These can be discouraged by scattering bits of tomato plants among the collards.

Comfrey
Symphytum officinale

Used for poultices, ointments for skin conditions and internally against arthritis, comfrey is a traditional medicinal plant. It accumulates potassium, calcium and phosphorus and has very high protein levels. Easy to grow in wet spots, it will extract nutrients from foul water and make these available for other plants after composting.

Conifers

Once established, these suppress weeds, and exudates from their leaves prevent seeds germinating. Pine needles are traditionally the best mulch for strawberries, improving both yield and flavour and discouraging slugs and snails. Protect young conifers from squirrels or rabbits by interplanting with members of the onion family.

Convolvulus tricolor

A hardy annual, convolvulus is pretty and resembles petunias but is easier to grow. It is an excellent hoverfly attractant, so it should be used widely under soft fruit and among vegetables.

Coriander
Coriandrum sativum

This repels aphids and has been used as a spray against spider mites. It helps anise germinate if the two are sown together.

Courgette (Zucchini)
Cucurbita pepo

These are marrow (squash) plants bred to produce lots of little fruits. They like the same conditions as the the other cucurbits - rich, warm and moist. Grow courgettes with sweet corn, peas or beans, which all get on well together: the legumes give nitrogen while the corn provides support and the courgettes keep the soil moist and shaded.

pea

54

courgette (zucchini)

day-lily

Day-Lily
Hemerocallis

These grow
almost anywhere;
the flowers and
buds are edible.
They go well
with irises in
sunny spots
where little else
will grow, but in
rich moist
conditions will
give a feast
of flowers.

Couch Grass

(Witch, twitch and quackgrass)
Agropyron repens

This common grass weed is suppressed by tomatoes, rape and the Mexican marigold, *Tagetes minuta,* which give off root secretions that poison and shade it out.

Cucumber

Cucumis sativus

Like courgettes, these are closely related to melons and squashes, so they need warm, rich, moist conditions. Ridge cucumbers do well under maize or sunflowers in their light shade. Similarly they like peas and beans, beet or carrots. Dill may aid the plants with root secretions and by attracting predators. Cucumbers trailing up and over sweet corn and beans make a happy trio.

Daffodil

Narcissus

These plants are believed to discourage mice as do other narcissi, grape hyacinths and scilla, so interplant them to hide crocus which mice are continually trying to eat.

Dill

Anethum graveolens

This annual herb is reputed to attract bees, though I find hoverflies and predatory wasps like it more. Dill aids cabbages and may help onions, sweet corn and cucumbers by repelling aphids and spider mites; it goes best of all among lettuces.

Eggplant

See aubergine

iris

57

Elm
Ulnus

Traditionally elms were grown for supporting grapevines as they were thought more beneficial than other trees.

Elderberry
Sambucus nigra

The berries are a good sacrificial crop to keep birds off fruit and a spray made from the leaves was once effectively used against aphids, carrot root fly, cucumber beetles, peach tree borers and root maggots. The presence of elderberry bushes assists composting; they also have a knack of seeding themselves by the bins. They are reputed to be good mole-deterrents.

grapevine

elm

Euphorbia
(Spurge family)

E. lathyrus or caper spurge
and *E. lactea* are among the
many plants people employ
hoping to drive away the
moles.

Ericaceae

These are a family of lime-haters or
calcifuges, of which heathers are among the
commonest. They give off leaf exudations
that prevent weeds germinating once the
heather is established.

Fat Hen
(Lamb's Quarters)
Chenopodium album

If it is not allowed to overwhelm,
this aids other plants, particularly
sweet corn and potatoes: its early
growth keeps soil moist. It is also
of benefit to the cucurbit or
marrow family and many flowers.
It can be used as a sacrificial plant,
attracting leaf-miners, and the
cool shade under its leaves is good
for ground beetles.

Feverfew

Matricaria or *Chrysanthemum*
parthenium

This belongs to the daisy family of
which many have insecticidal
properties but this plant has a strong
scent which can hide others from
pests. The golden form makes a
particularly attractive edging and
retains its cheerful colour in sun or
shade. It self-seeds with a vengeance
but is easy to hoe or weed. If sown in
situ, it will even thrive in tough spots
such as in front of dry hedges.

feverfew

Flax
Linum usitatissimum

Used as a green manure or as an intercrop
it benefits carrots and potatoes by
improving the soil tilth. It will hide the
potatoes from Colorado beetles. It is very
beautiful flowering *en masse* and is much
loved by bees.

Fig
Ficus

Surprisingly hardy and tough plants, these
attractive shrubs will grow in any sunny
spot. They do well with rue, which covers
their bare stems, and their colours look
attractive together.

Foxglove
Digitalis

Poisonous but rarely causing any trouble, its presence stimulates the growth of plants, especially pine trees and rhododendrons. The plants accumulate iron, calcium, silica and manganese to high levels in the leaves. It is widely believed that foxgloves are generally beneficial to all parts of the garden, so they should be encouraged to self-sow freely. A tea made from the leaves may be used to keep cut flowers fresh.

pine

foxglove

Garlic
Allium sativum

The most pungent of the onion family, garlic is a highly effective accumulator of sulphur, which may explain its very ancient reputation as a fungicide. Garlic emulsion kills aphids and onion flies; it has also been used against codling moths, snails, root maggots, Japanese beetles, carrot root fly and peach leaf curl. The cloves put in with stored grain will discourage weevils. Garlic is not as attractive as chives, but it is especially good for roses and fruit trees.

gooseberry

Gooseberry
Ribes grossularia

These are less prone to mildew
when grown on a stem, although
they then need staking. This
allows underplanting with
Limnanthes douglassii, the poached
egg plant, which is very beneficial.
Tomatoes may aid them, and
interplanting broad beans may
drive away the sawfly caterpillar.

poached egg plant

Geranium
Pelargonium

The tender houseplant, not the wild cranes-
bills, has very strongly scented foliage and
versions with lemon or ginger overtones can be
used to hide other plants. The white-flowered
form is said to be good for maize, and may have
been used against cabbage worms and beetles.

Grapevine
Vitis vinifera

Grow grapes over elm or mulberry trees if
you wish to do as the Romans did. Grape-
vines are aided by blackberries which
improve the soil and provide other fruits for
the birds, as does asparagus. Hyssop and
mustard are beneficial to grapes and clover
should be included in the mixture when the
trees are put down to grass.

Hazel
Corylus avellana

The cultivated forms such as Cosford cob or
Webb's prize cob produce bigger nuts than the
wild forms, yet support as many varieties of
wildlife. They are beneficial in hedges and
pastures for fodder and as fly deterrents. They
make excellent wind-breaks and good screens for
compost bins or stables.

Hemp
Cannabis sativa

This may increase the flavour and keeping qualities of vegetables and fruit. It was once used extensively to deter cabbage-white butterfllies, which is no longer practical as it is unlawful to cultivate it in most countries. It has bactericidal qualities and may also be fungicidal; it is certainly reputed to deter potato blight.

Hop
Humulus

These climbers suffer from wireworms, which can be lured on to a trap planting of potatoes grown between the rows.

Horehound
Marrubium vulgare

A wild flower of chalky places. It was traditionally used for medicinal purposes. It stimulates and aids fruiting in tomatoes.

Horseradish
Armoracia rusticana

This aids potatoes but as it is almost impossible to get rid of horseradish, it i best grown in large pots so it can't escape. Horseradish accumulates calcium, potassium and sulphur. In the USA it has been used against blister beetles and Colorado beetles. The tea has been used to prevent brown rot in apples, so grow it under the trees.

potato

Hyssop
Hyssopus officinalis

This superb bee plant benefits grape-vines and makes a good ground cover underneath them. Grown by the vegetable bed it helps ward off the cabbage butterflies.

Kohlrabi
Brassica oleracea v.
caulorapa

A tough and disease-resistant crop much like a turnip, but more drought tolerant, and not as hot to the taste. It does well intercropped with onions and beet.

hop

Lavender
Lavandula officinalis

Always used to keep pests out of clothes, it also keeps them away from the garden. Grow cotton lavender (*Santolina incana*) and Russian lavender (Perovskias), which are unrelated but beneficial, especially to bees and other friends. *Lavandula stoechas* or French lavender is wonderfully aromatic but not very hardy. All the lavenders are good for lilies, supporting and sheltering their lower stems.

onion

Lamb's Quarters

See fat hen

Legumes

These are the plants that fix nitrogen from the air by means of nodules on their roots. The roots harbour bacteria and fungi, which do the fixation in return for nutrients from the plant. As well as the clovers, peas, beans and lupins we usually think of, trees such as laburnum, the Judas tree (*Cercis siliquastrum*) and caragana, and shrubs such as the brooms and tree lupins are also leguminous. All of them benefit other plants by supplying nitrogen as their root hairs and associated micro-organisms decompose.

carrot

Leeks
Allium porrum

Another allium, this likes onions and goes well with celery in deep, rich soil. Grow carrots, onions and leeks all intermingled to decrease attacks of carrot and onion root flies which are confused by the smells. Leeks also hide the brassicas from pigeons.

Lettuce
Lactuca sativa

Sow a little and often. These do best among cucumbers, carrots, radishes and strawberries, as they create a humid environment that all enjoy. Lettuce can be protected from aphids with chervil and dill, which always does well in among the lettuces and attracts many predatory insects. If it is hot, sow lettuce on the shady side of these companions, as they will not germinate in temperatures above 65°F (18°C). Lettuce is one of the plants that has been shown to take up natural antibiotics from the soil which may then be of use to other plants, and us.

Lime
Tilia

These trees get very big, so their insignificant flowers which are full of nectar add up and one tree may fill a hive with honey.

Lovage
Levisticum officinale

This herb adds 'body' to vegetable stocks and substitutes for salt. A big plant, it likes moist soil and aids most other plants by its presence, as the damp shade under its leaves provides a good habitat for ground beetles.

lupin

Lupin
Lupinus

These add nitrogen, so aiding sweet corn,
and they are very useful green manure,
improving soil texture. They may support
vast populations of aphids and thus predators
after flowering. They have been noted since
classical times for their ability to suppress
weeds, especially fat hen. Fill up gaps in
beds and borders with lupins and they can
be just as beneficial in among the vegetables.

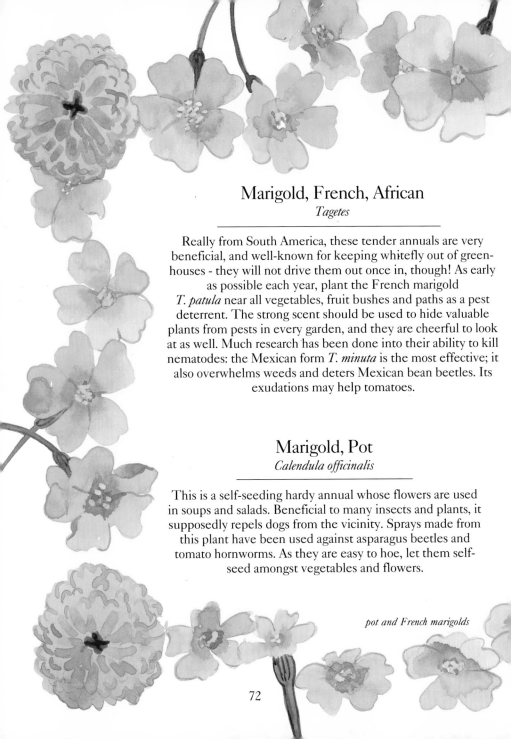

Marigold, French, African
Tagetes

Really from South America, these tender annuals are very beneficial, and well-known for keeping whitefly out of greenhouses - they will not drive them out once in, though! As early as possible each year, plant the French marigold *T. patula* near all vegetables, fruit bushes and paths as a pest deterrent. The strong scent should be used to hide valuable plants from pests in every garden, and they are cheerful to look at as well. Much research has been done into their ability to kill nematodes: the Mexican form *T. minuta* is the most effective; it also overwhelms weeds and deters Mexican bean beetles. Its exudations may help tomatoes.

Marigold, Pot
Calendula officinalis

This is a self-seeding hardy annual whose flowers are used in soups and salads. Beneficial to many insects and plants, it supposedly repels dogs from the vicinity. Sprays made from this plant have been used against asparagus beetles and tomato hornworms. As they are easy to hoe, let them self-seed amongst vegetables and flowers.

pot and French marigolds

72

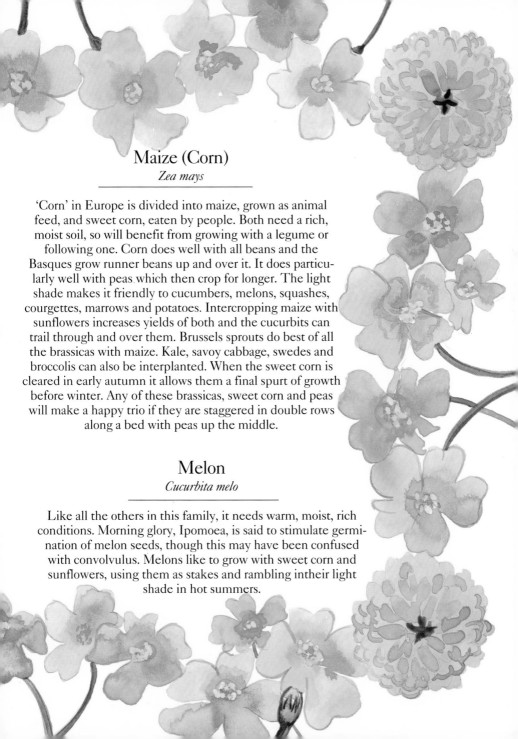

Maize (Corn)
Zea mays

'Corn' in Europe is divided into maize, grown as animal
feed, and sweet corn, eaten by people. Both need a rich,
moist soil, so will benefit from growing with a legume or
following one. Corn does well with all beans and the
Basques grow runner beans up and over it. It does particularly well with peas which then crop for longer. The light
shade makes it friendly to cucumbers, melons, squashes,
courgettes, marrows and potatoes. Intercropping maize with
sunflowers increases yields of both and the cucurbits can
trail through and over them. Brussels sprouts do best of all
the brassicas with maize. Kale, savoy cabbage, swedes and
broccolis can also be interplanted. When the sweet corn is
cleared in early autumn it allows them a final spurt of growth
before winter. Any of these brassicas, sweet corn and peas
will make a happy trio if they are staggered in double rows
along a bed with peas up the middle.

Melon
Cucurbita melo

Like all the others in this family, it needs warm, moist, rich
conditions. Morning glory, Ipomoea, is said to stimulate germination of melon seeds, though this may have been confused
with convolvulus. Melons like to grow with sweet corn and
sunflowers, using them as stakes and rambling in their light
shade in hot summers.

Mineral-accumulating Plants

Weeds are often the best at this, as they thrive on soils that are
short in those very elements. They can be made into teas, liquid
feeds or sprays but it is best to use them for compost which can
return the scarce minerals to the crops that need it most. Pernicious weeds such as couchgrass can be withered before
composting; weeds in seed should only be put in the
middle of a heap you expect to get hot, where they will
be destroyed. The following is a list of some of the
minerals plants need:

Boron is needed in minute amounts, particularly by
brassicas and apples; it is accumulated by the euphorbia
or spurge family.

Calcium is found in the soil as lime. Earthworm activity makes it
more available, so feed them with seaweed and grass clippings.
Many plants such as rhubarb accumulate calcium as poisonous
crystals of oxalate to deter grazing.

Lime is accumulated by beech, brassicas, broom, buckwheat, cacti,
coltsfoot, comfrey, corn chamomile, corn marigold, creeping
thistle, daisy, dandelion, equisetum, fat hen, goosegrass, melons,
oak leaves and bark, okra, purslane, robinia, scarlet pimpernel,
shepherd's purse, silverweed and stinging nettle.

Cobalt is one of the rarest elements and is essential to animals but
hardly needed by plants. It is concentrated by buttercup, comfrey,
equisetum, ribbed plantain and rosebay willow herb.

Copper is often added to pig food, so it is rather too readily
supplied in pig muck; plants growing on this may be poisoned.
Generally in short supply, it is accumulated by buttercup, chickweed, coltsfoot, creeping thistle, dandelion, plantain, sow and
spear thistle, stinging nettle, vetch and yarrow.

Iron: a shortage of this, or too much lime locking it up, causes
yellowing growth and other symptoms of chlorosis. Iron can be
obtained from beans, buttercup, chickweed, chicory, coltsfoot,
comfrey, creeping thistle, dandelion, equisetum, fat hen, foxglove,
ground ivy, groundsel, silverweed (*Potentilla anserina*)
and stinging nettle.

Magnesium is another element that can be locked up by lime,
causing chlorosis. It is found in quantities in beech,
chicory, coltsfoot, daisy,
equisetum, larch, oak,

dandelion

potatoes, plantain, salad burnet, silverweed and yarrow.
Manganese can also be locked up by lime. The beet family
need it more than other crops. It is accumulated by
buttercup, chickweed and comfrey.

Nitrogen is accumulated in abundance by the following
plants but only while they are young and succulent: beans,
bindweed, black nightshade, broad-leaved dock, chickweed, clovers, comfrey, creeping thistle, dandelion, fat
hen, grass clippings, groundsel, knotgrass, peas, purslane,
sow thistle, stinging nettle, thorn-apple, vetches, white
campion and yarrow. Throughout its life, any legume fixes
surplus nitrogen, mostly in the root system.

Phosphorus is needed for good roots. This used to be
provided by the use of bone-meal fertilizer. Seaweed
products are pleasanter but contain less phosphorus. Plants
accumulating it are broad-leaf dock, buttercup, comfrey, fat
hen, henbane, oak leaves, purslane, sheep's sorrel, vetch
and yarrow.

Potassium is needed for fruiting and disease-resistance,
most of all by gooseberries and cooking apples. It is found
in apple leaves and twigs, beech leaves, broad-
leaved dock, buttercup, chickweed, chicory,
coltsfoot, comfrey, corn chamomile, couchgrass,
fat hen, goosegrass, maple leaves, plantain,
purslane, stinging nettle, sunflower, tansy, thistles,
thorn-apple, tobacco, vetch and yarrow.

Silica makes plants tougher and more disease-resistant.
It is most plentiful in equisetum - mare's-tail or horsetail
but is also found in couchgrass, horseradish, knotgrass,
onions, plantain and stinging nettle.

Sulphur is important for disease-resistance. It is
found in the Allium family and in brassicas,
coltsfoot, fat hen and purslane.

dock

walnut

Mints
Mentha

Plant mints in pots buried in the soil to minimize their expansionist tendency. One of the few plants to grow under walnuts, mint thrives near stinging nettles and helps cabbage and tomatoes before it overwhelms them. It can be used to repel rodents, fleas and flea beetles, and spearmint discourages aphids by discouraging their 'owners', the ants. Sprays of mint tea repel ants and Colorado beetle. Mints are wonderful autumn bee plants; they also feed hoverflies and predatory wasps, so they make a cheap and effective ground cover under established fruit trees.

Morning Glory
Ipomoea

Traditionally grown by the Indians with their sweet corn to give it vigour, morning glory is also thought to aid germination of melon seed. As these tender crops grow well together they make a happy trio. Start them off in pots indoors and then plant them out after hardening off; put the sweet corn out first to provide shelter.

mint

Mugwort
Artemisia vulgaris

A common roadside weed closely related to wormwood, this has the exudates which repel insects. Mugwort is said to be liked by hens; it may help control their lice and worms.

Mulberry
Morus nigra

One of the best trees to support grapes, mulberries are traditionally grassed down underneath and provided with a circular seat.

Mustard
Brassica alba

What is often called mustard is actually oilseed rape; true mustard is not as hardy and is killed by hard frosts. It makes a good cover crop, leaving soil in fine condition. Mustard can be used to attract many pests of the Brassica family, then dug in or composted. It should be used with legumes when grassing down an orchard for ease of maintenance and to prevent soil erosion; it is a corrective green manure on acid soils.

Nasturtium
Tropaeolum majus

Nasturtiums can be used to enliven salads and do best on poor soil. Their strong smell drives woolly aphis off apple trees and keeps aphids away from broccoli and squash. They are themselves attacked by black aphids and cabbage caterpillars, so can be used as sacrificial crops. They keep whitefly out of greenhouses and are helpful rambling among radishes, potatoes, brassicas and cucurbits.

stinging nettles

Nettle, dead
Lamium album

A nectar-producer over a long season, which makes it very valuable to early bees and other insects, this weed should be encouraged. As dead-nettles thrive in damp, shady spots and are generally beneficial to vegetables, they can be allowed to flourish in drainage channels and ditches. They may also help deter potato bugs.

Nettle, stinging
Urtica dioica

While dead-nettles help pollinators or predators, stinging nettles are of direct benefit to other plants. The soil they leave is rich in humus and iron and they are very good ground cover. A spray made from stinging nettles is rich in calcium and silica, useful for invigorating plants, and either alone or mixed with comfrey they rot down to make an excellent liquid feed. Encourage them in hedges and wild areas; patches of nettles in orchards and paddocks are also beneficial unless they are allowed to run rampant.

dead nettles

Onion
Allium cepa

Brassicas do well with onions, as do beet, tomatoes and lettuce; they all benefit from the smell of onions hiding them from pests. Strawberries are less vulnerable to disease when grown with onions, which enjoy the open situation. Summer savory may help the onions, but it also helps beans which do not get on with them. Chamomile will aid in small amounts. Carrots are often intercropped so that the onion and carrot flies are both confused. This is even better when they are grown in a trio with leeks; early onions and late leeks do best. Onions deter Colorado beetles and rabbits, though hungry ones will ignore them. When the onions are ripening it is beneficial to let the weeds grow; they take up moisture and nutrients, particularly nitrogen, and the bulbs keep better.

Okra
Hibiscus esculentus

One of the less hardy crops as it needs warm, rich, moist conditions, as do melons and cucumbers with which it grows well. Plant the three together and let the melons and cucumbers trail over the ground while the okra stands above. Alternatively, grow these climbers over sweet corn and sunflower plants with the okra on the sunny side.

parsley

Parsley
Petroselinum crispum

A health-giving herb that loves rich, moist soil,
parsley is difficult to establish or move, so let it
self-seed where it will. It aids tomatoes and
asparagus, making a happy trio if enough water is
present. It has often been used to mask carrots
and onions from their root flies, though it can
suffer from carrot fly itself. It has been used as a
tea to ward off asparagus beetles, and it attracts
hoverflies.

Parsnip
Pastinaca sativa

Parsnips are very slow to germinate, so sow them with radishes which will come up quickly and mark the positions. They grow well with lettuce and peas if not shaded too much. The flowers attract hoverflies and predatory wasps.

Pea
Pisum sativum

These grow well with root crops, all of the beans, potatoes, cucurbits and sweet-corn. Growing dwarf varieties in between other vegetables is always of benefit, except with the alliums.

Peppermint
Mentha piperita

The oil is much used for scent and flavour and its production is improved if you grow stinging nettles nearby. In the presence of chamomile, peppermint produces less oil, but chamomile more. Like most of the mint family it will help keep brassicas free of cabbage-white butterfly caterpillars by disguising their scent. This mint is especially loved by bees and other beneficial insects.

Phacelia tanacetifolia

A pretty plant useful as ground cover and green manure, this also attracts many beneficial insects, especially hoverflies. It is very useful sown between shrubs or roses, as it will keep the soil covered and is easy to remove or incorporate when finished.

pea

Pepper, sweet
Capsicum

This is a tender plant very similar to hot pepper but without the burning effect; it is prone to aphids. It likes basil, which grows well with it, and they are both usually grown with okra for convenience, as they all need warmth and shelter.

sweet peppers

Pepper, hot
Capsicum

These tender plants are prone to aphids but a spray of hot chilli pepper made from their dried fruits and seeds is effective at discouraging many pests. They like much the same treatment and conditions as sweet peppers and grow well with basil, which they closely resemble when not in fruit. Some chillis have root exudates that inhibit root rot and other Fusarium diseases, so they may be beneficial to many other plants.

83

Poached egg plant
Limnanthes douglassii

Low-growing annual with ferny foliage which
makes a good ground cover, it self-seeds pro-
fusely. It is one of the very best hoverfly attractants and is also liked by bees. Grow it under
shrubs, roses and in the fruit cage. As it is so low-
growing it does not trouble the bushes and is one
of the best all-round companions.

Potato
Solanum tuberosum

Being highly bred and much grown, potatoes are very prone to pests and disease.
Horseradish is a vigorous weed that is often
planted to aid them but must be carefully
contained. Peas help them considerably.
Tagetes marigolds are beneficial, as are
celery, flax, Lamium, nasturtium and savory
which all help protect them from pests.
Potatoes are mutually beneficial with beans,
maize and cabbages. In the USA aubergines
(egg plants) are used as sacrificial crops to
the Colorado beetle. Potatoes make a happy
trio with sweet corn and peas or beans. You
can protect them from scab by putting grass
clippings and comfrey leaves in when
planting the sets.

raspberry

tansy

Pumpkin
Cucurbit

This has much the same likes
and dislikes as other cucurbits
but needs immense fertility and
copious water to grow really big.
Pumpkins may be healthier
with Datura (thorn-apple)
weeds nearby and they do well
in a happy trio under sweet corn
and sunflowers.

Radish
Raphanus sativus

Although it is normally grown for its
roots, the small seed pods are delicious.
More than most crops, radishes need to
be grown rapidly to be edible. As they
are closely related to cabbages and other
brassicas, they should usually be kept
apart, though in seed-beds they attract
pests such as flea beetles to themselves.
Radishes get on with most other plants
and benefit from the presence of
nasturtium and mustard. Chervil,
lettuce, peas and radishes will all get on
well together.

Raspberry
Rubus

These woodland plants need
rich, moist, well-mulched soil
and benefit from being underplanted
with tansy and marigolds of any low-
growing variety. Strawberries will
happily grow in the light shade
between their rows.

Rhubarb
Rheum rhaponticum

Boiled rhubarb leaves were used in ancient times as an aphicide, as they contain the poison oxalic acid. The same spray was traditionally used against black spot on roses. The plant may help control red spider mites on aquilegias (columbines). Including bits of rhubarb in the sowing or planting holes used to be a method of controlling clubroot in brassicas.

Rosemary
Rosmarinus

This perennial but not very hardy herb is one of the most valuable to humans, bees, insects and other plants. Its exudations inhibit seeds germinating underneath it. Plant it in as many sheltered spots as you can squeeze it in, near but not next to flowers, fruits and vegetables.

Rue
Ruta gravaeolens

Rue likes figs and can be planted underneath to cover the bare stems. The smell will drive away fleas and Japanese beetles from homes and gardens. Generally a good influence on other plants and although it is poisonous it used to be grown near stables to keep away flies.

Rose
Rosa

Roses are grown everywhere, so they are very prone to pests and diseases. Underplant them with alliums, especially chives and garlic, which prevent bad attacks of black spot and increase their perfume. Parsley, thyme, catnip, mignonette, lupins and *Limnanthes douglassii* are also beneficial.

rose

allium

Sage
Salvia officinalis

Sage benefits the brassicas and carrots because its smell confuses their pests. It gets on particularly well with rosemary and is very beneficial to plants and insects, so plant it near the vegetables.

Salsify
Trapopogon porrifolius

This long, thin, carrot-like vegetable is very similar to scorzonera (see below). A good plant to grow with water-melons and mustard, it can be used to discourage carrot root fly. The attractive flowers are beneficial to insects of many kinds.

Savory
Satureia

One of the savories is useful against Mexican bean beetles, although it is not known whether it is the winter or summer variety. But both repel many insects. Summer savory, *S. hortensis*, has the better flavour and is very good eaten with broad beans; grown with them it will keep them free of black aphids. Similarly it will go well and grow well with onions, but these will not make a trio as the onions do not get on with the beans.

sage

Scorzonera hispanica

Similar to salsify with long, thin, carrot-like roots, it will also help repel carrot root fly. The attractive flowers are beneficial to insects in the same way as salsify but they are purple-blue instead of yellow. Both plants can usefully be included in either the flower or the vegetable garden.

Shallot
Allium cepa/ascalonicum

Grow these under fruit trees, especially apples, which then suffer less scab. They can also be grown to advantage between strawberry plants, though some people think this looks odd.

Southernwood
Artemisia abrotanum

This is a relative of wormwood that discourages moths and insects and has a lovely scent of lemony pine. It can be used in borders as a pest deterrent, particularly for brassicas and carrots. Plant it by corners and gates where you will brush against it often to release the scent.

rosemary

Sunflower

Helianthus annuus

The flowers are good for bees, lace-wings and predatory wasps, while the seeds are much beloved of wild birds and chickens as well as for human snacks. Hungry feeders, sunflowers can be used as green manure on rich soil, but must be incorporated or composted before they get too tough. They get on well with sweet corn and are mutually protective, as well as helping break the wind. A happy trio can be made with cucumbers and other cucurbits such as melons or courgettes growing underneath and climbing over these natural stakes.

sunflower

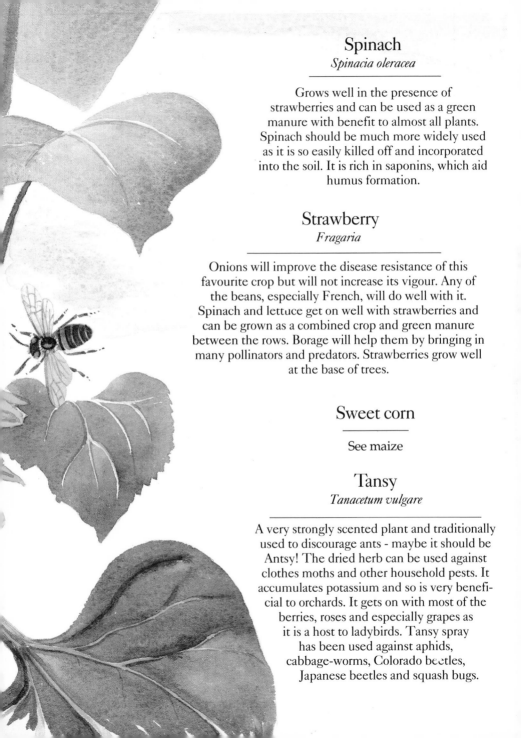

Spinach
Spinacia oleracea

Grows well in the presence of strawberries and can be used as a green manure with benefit to almost all plants. Spinach should be much more widely used as it is so easily killed off and incorporated into the soil. It is rich in saponins, which aid humus formation.

Strawberry
Fragaria

Onions will improve the disease resistance of this favourite crop but will not increase its vigour. Any of the beans, especially French, will do well with it. Spinach and lettuce get on well with strawberries and can be grown as a combined crop and green manure between the rows. Borage will help them by bringing in many pollinators and predators. Strawberries grow well at the base of trees.

Sweet corn

See maize

Tansy
Tanacetum vulgare

A very strongly scented plant and traditionally used to discourage ants - maybe it should be Antsy! The dried herb can be used against clothes moths and other household pests. It accumulates potassium and so is very beneficial to orchards. It gets on with most of the berries, roses and especially grapes as it is a host to ladybirds. Tansy spray has been used against aphids, cabbage-worms, Colorado beetles, Japanese beetles and squash bugs.

Turnip
Brassica rapa

Another of the cabbage family with many of their traits, turnips get on very well with peas and are benefited by hairy tare (*Vicia hirsuta*), which will keep off aphids. Flea beetle is a problem best cleared simply by spreading bits of tomato leaf which drives them away.

Tomato
Lyco persicon esculentum

Tomatoes have been said to aid early cabbage - which seems unlikely as their growing seasons hardly touch. However, bits of the leaf do repel flea beetles. They get on well with basil, alliums, parsley, marigolds and nasturtiums, and particularly well with asparagus. Tomato-leaf spray has been used against asparagus beetles and the asparagus roots kill trichodorus, a nematode that attacks tomatoes.
So plant tomatoes and asparagus in a happy trio with basil or parsley. A strange but effective combination is to plant tomatoes near roses, protecting them from black spot; similarly they may be protective to gooseberries.

Valerian
Valeriana officinalis

One of the most generally helpful and beneficial plants, valerian is good for insect friends, cats and earthworms. It stimulates composting and accumulates phosphorus. The red valerian (*Centranthus ruber*) may be equally valuable.

Wallflower
Cheiranthus cheiri

Apart from being one of the easiest perfumed plants to bring spring cheer, wallflowers benefit apple trees; this is probably because they are early flowering and thus bring in more pollinators.

Yarrow
Achillea millefolium

Yarrow helps oil production and vigour in other herbs and plants. It accumulates phosphorus, calcium an silica, so it is good for the compost heap. Yarrow is a very good and long flowering host to hoverflies, ladybird and predatory wasps.

Zucchini

See courgette

Good and Not so Good Companion
Guide to the Vegetable Patch

Crop	Good Companions	Not so good Companions
Asparagus	Basil, parsley, tomato	
Aubergine	Beans	
Beans, broad or field	Brassicas, carrot, celery and celeriac, cucurbits, potato, summer savory, most herbs	Onions and garlic
Beans, French	Celerys, cucurbits, potato, strawberry, sweet corn	Onions and garlic
Beans, runner	Sweet corn, summer savory	Beetroot and chards, kohlrabi and sunflower
Beetroot and chards	Most beans, brassicas, onions and garlic, kohlrabi	Runner beans
Brassicas	Beetroot and chards, celery and celeriacs, dill, nasturtium, onions and garlic, peas, potato	Runner beans, strawberry, tomato
Carrot	Chives, leek, lettuce, onions and garlic, peas, tomato	Dill
Celery and celeriac	Brassicas, beans, leek, tomato	
Cucurbits	Beans, nasturtium, peas, radish, sunflower, sweet corn	Potato
Kohlrabi	Beetroot and chards, cucurbit	Potato, tomato, runner beans
Leek	Carrot, celery and celeriac, onions and garlic	
Lettuce	Carrot, cucurbits, radish	
Onion and Garlic	Beetroot and chards, lettuce, strawberry, summer savory	Beans, peas
Pea	Beans, carrot, cucurbits, radish, sweet corn, turnip	Onions and garlic
Potato	Beans, brassicas, peas, sweet corn	Cucurbits, sunflower, tomato
Radish	Chervil, cucurbits, lettuce	
Spinach	Almost everything	
Strawberry	Beans, lettuce, spinach	Brassicas
Sunflower	Cucurbits, sweet corn	Potato
Sweet corn	Beans, cucurbits, peas, potato, sunflower	
Tomato	Asparagus, basil, carrot, onions and garlic, parsley	Brassicas, kohlrabi, potato
Turnip and swede	Peas	

Index

figures in bold denote main entries